Cooling

by Sheila Rivera

first step nonfiction

Lerner Publications · Minneapolis

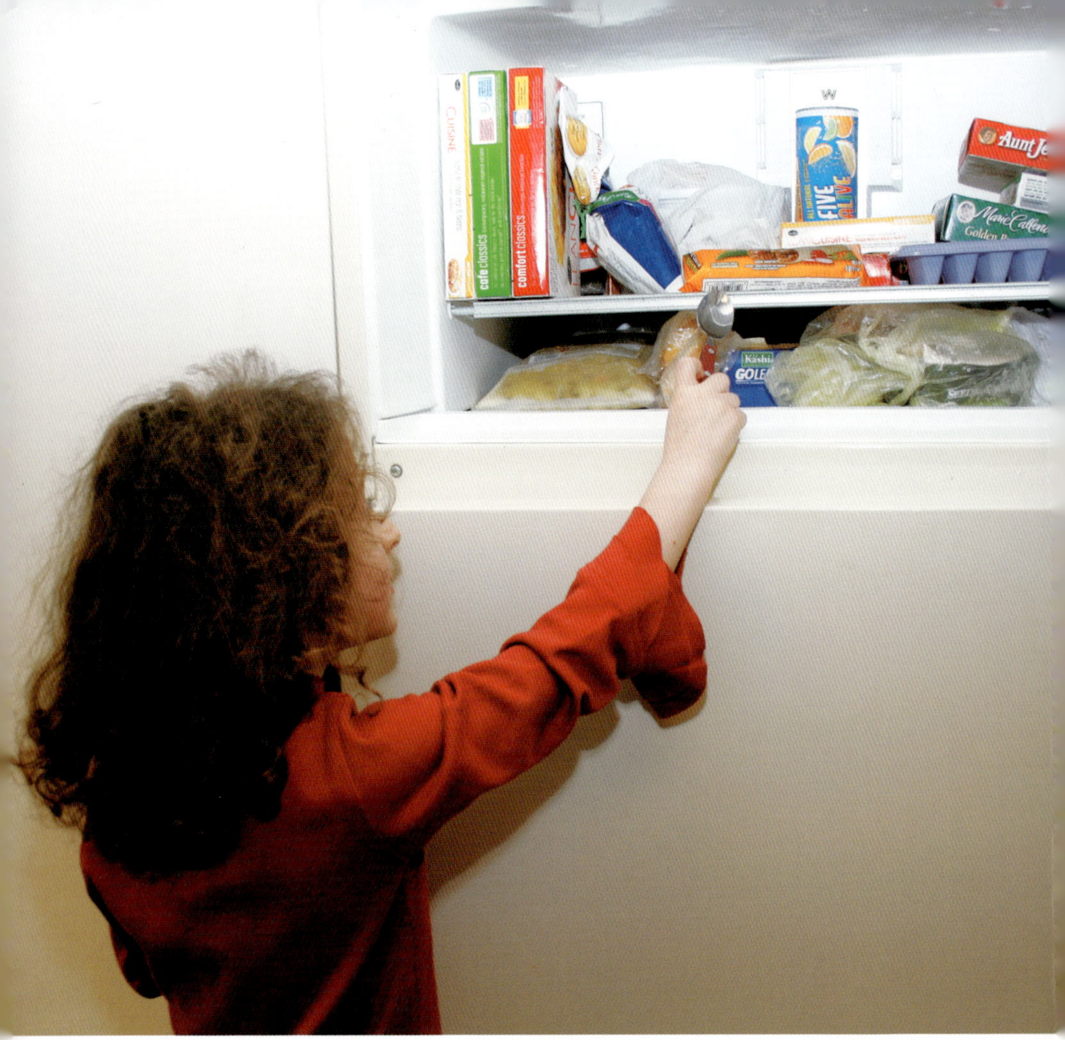

I can cool a spoon.

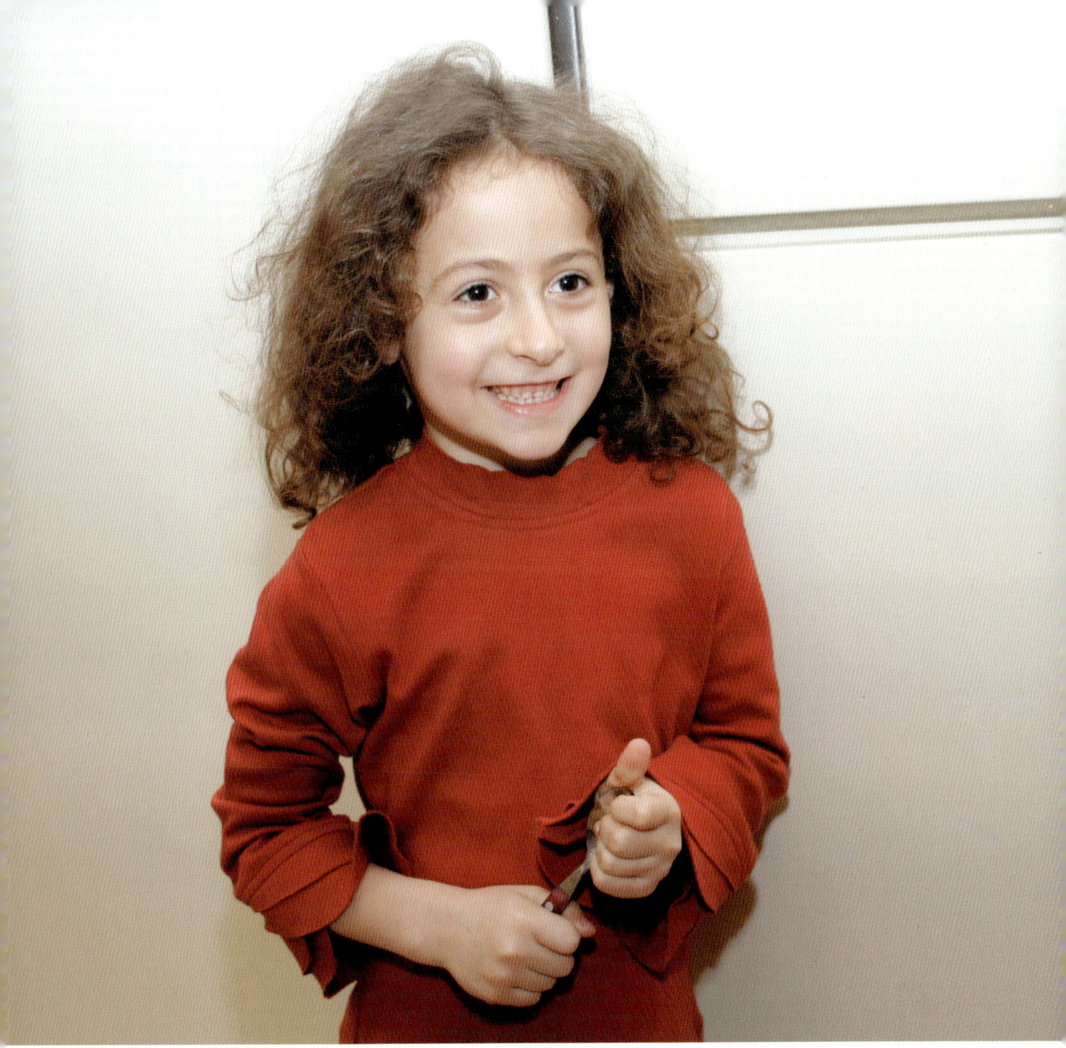

It gets cold.

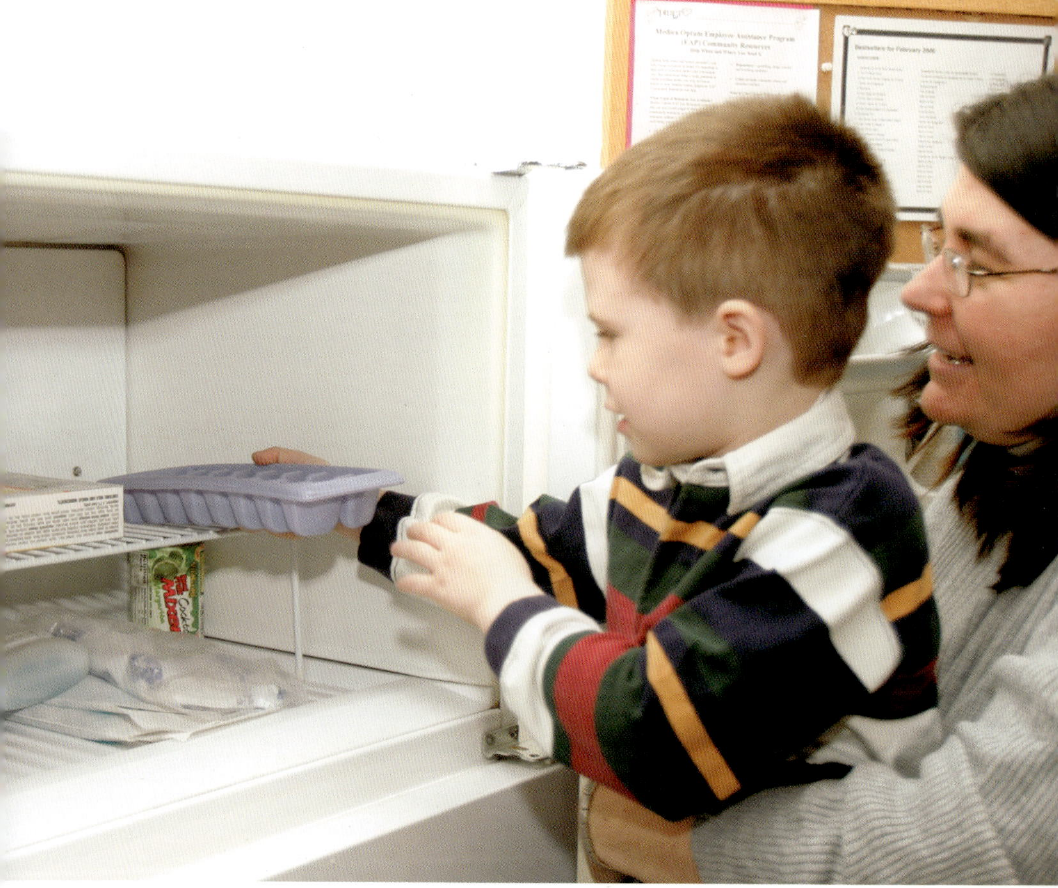

I can cool water.

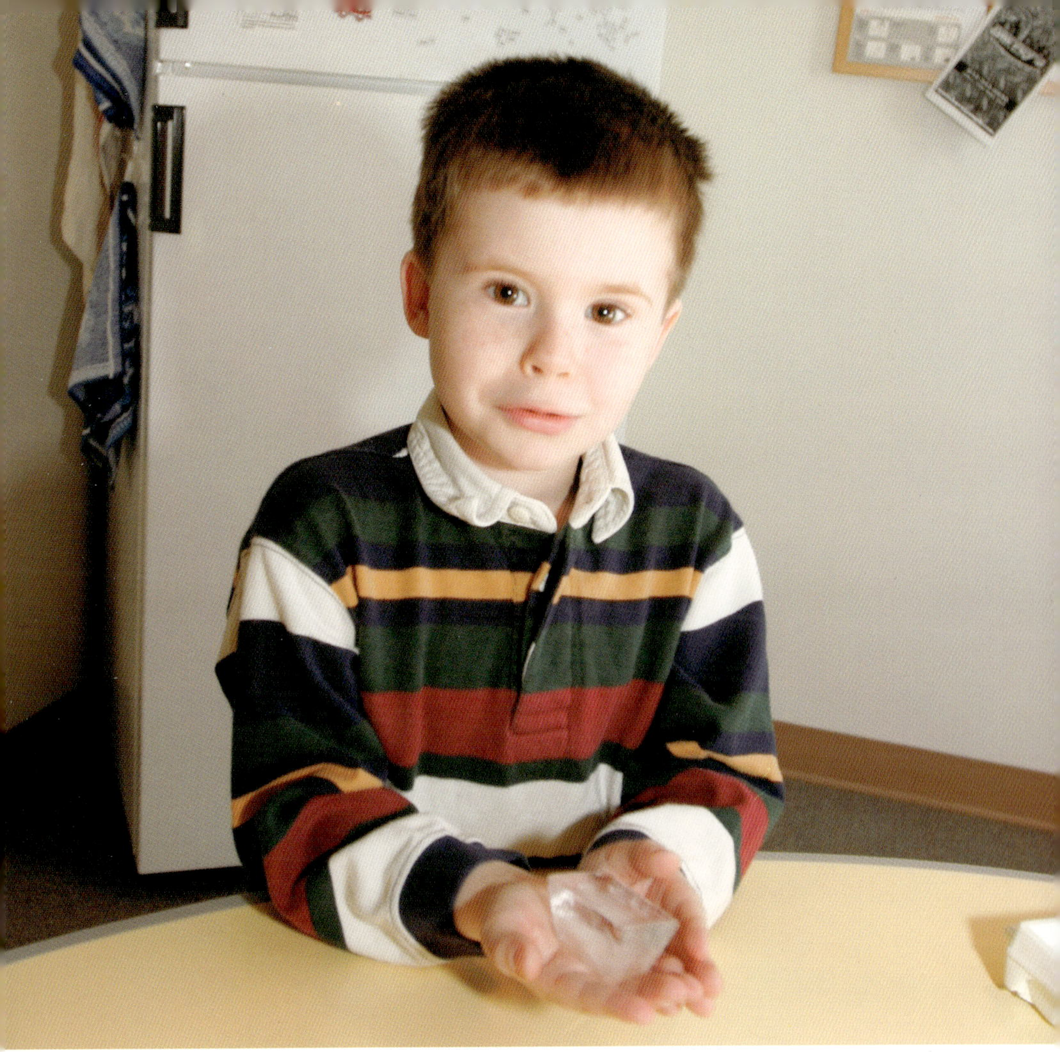

It freezes.

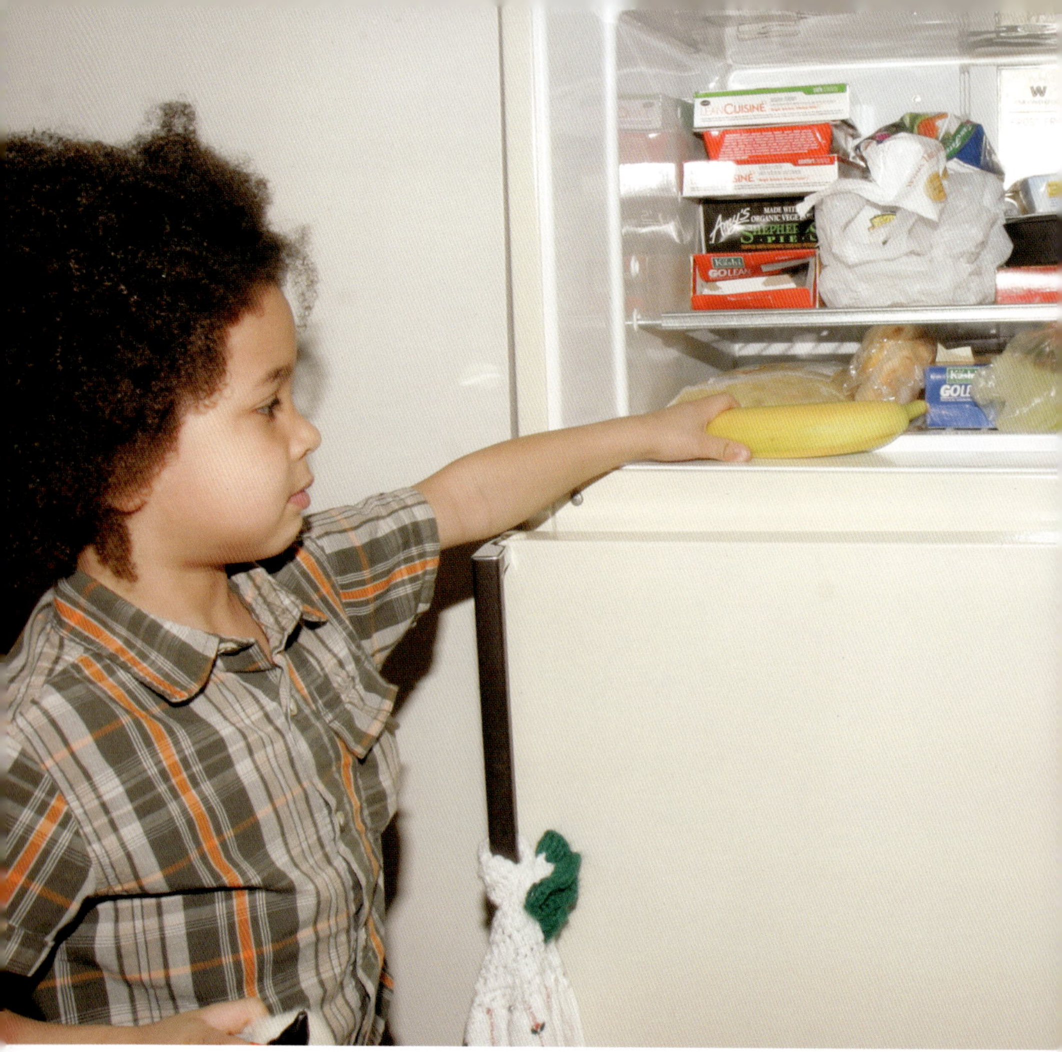

I can cool a banana.

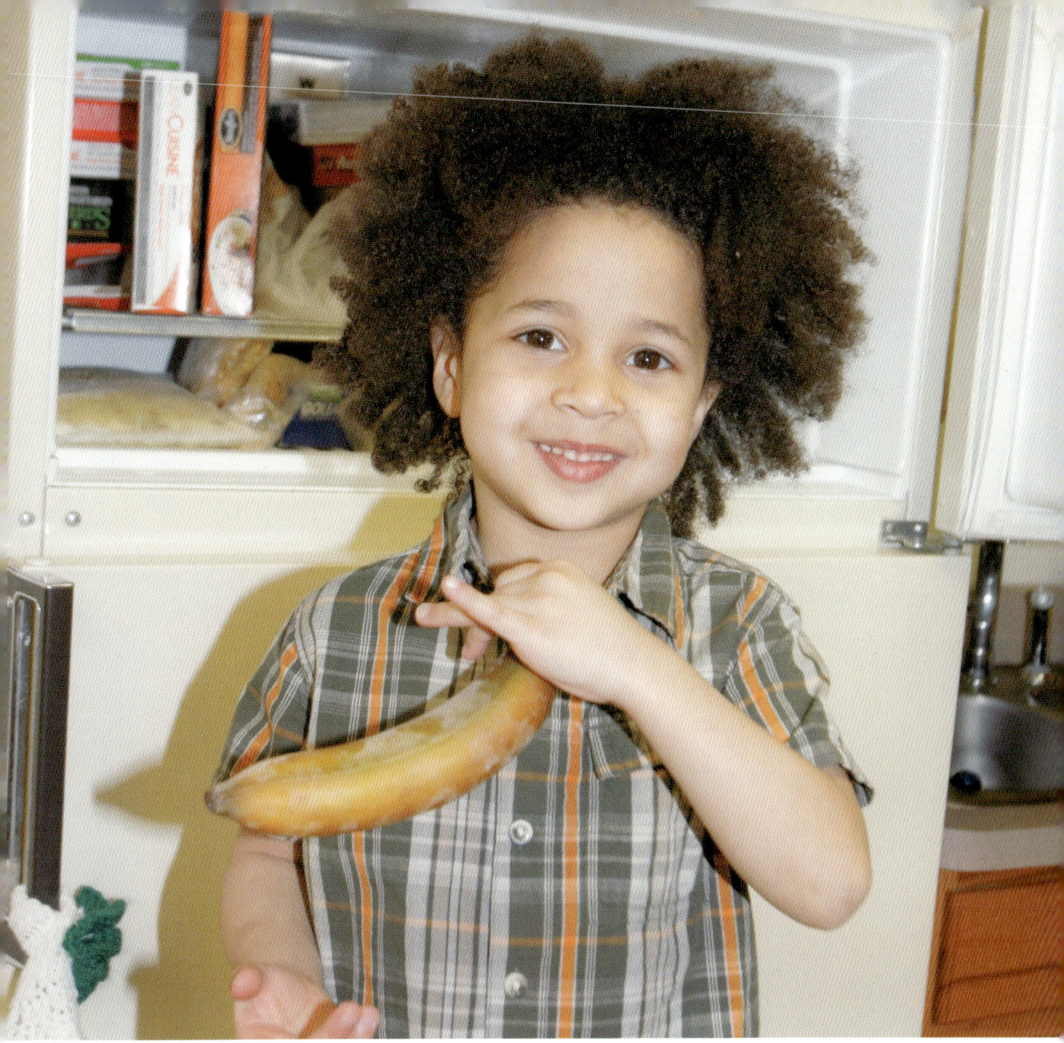

It changes color.

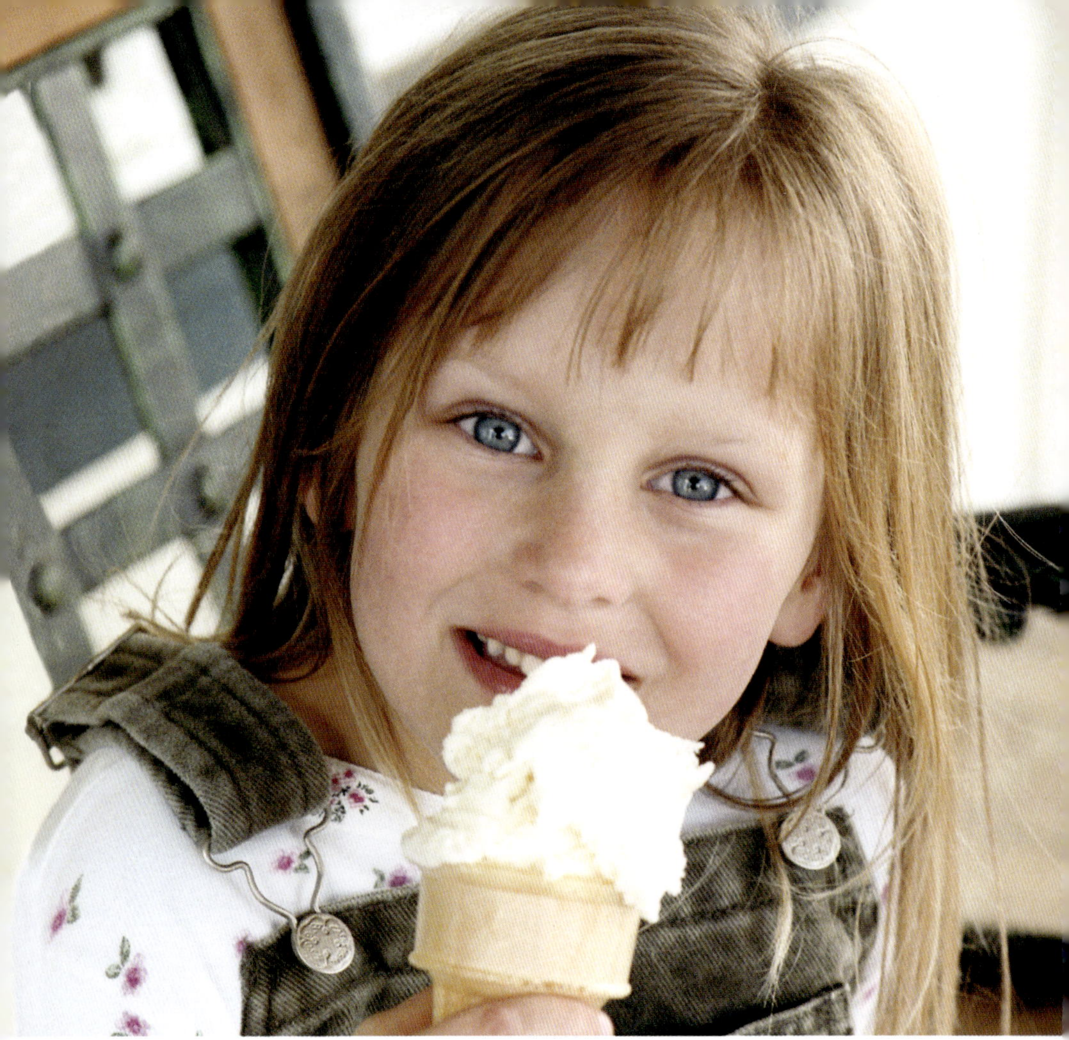

What can you cool?